junior Charles Catton

Animals Drawn From Nature And Engraved In Aqua Tinta

junior Charles Catton

Animals Drawn From Nature And Engraved In Aqua Tinta

ISBN/EAN: 9783744739924

Printed in Europe, USA, Canada, Australia, Japan

Cover: Foto ©berggeist007 / pixelio.de

More available books at **www.hansebooks.com**

ANIMALS

DRAWN FROM NATURE,

AND

ENGRAVED IN AQUA-TINTA,

BY

CHARLES CATTON, Jun.

WITH A DESCRIPTION OF EACH ANIMAL.

LONDON:

Printed for the AUTHOR, No. 7, on the Terrace, Tottenham-Court Road; and sold by I. and J. TAYLOR, at the Architectural Library, No. 56, Holborn, opposite Turn-stile.

M.DCC.LXXXVIII.

TABLE OF CONTENTS,

AND ORDER IN WHICH TO ARRANGE THE PLATES.

THE LION

Publ. by Edw. Orme, London, February 1, 1807.

The L I O N,

BEING universally esteemed king of the forest, deserves a pre-eminence of station: we therefore introduce him first to notice.

The Lion is classed by the Naturalists in the Cat Tribe; 'tis an animal, of all others the most majestic and stately; with a large head, the upper part of which, with the chin, the whole of the neck and shoulders, are clothed with a long shaggy hair, resembling a mane; the hair of the body and limbs is short and smooth, but long on the bottom part of the belly; a long tail, which appears of equal thickness by reason of the increasing length of hair towards the end, where it terminates in a large black tuft: the colour is tawny, or dirty brown; the belly part inclines to white. The limbs of this animal are of vast strength, and need only be seen to convince of their force. The country where they most abound is Africa, the most wild and desert parts: they also are found in parts of Asia, but the former appears most congenial to the Lion's constitution, those of that country being much larger (having been known 5 feet high, and 10 feet long), and more fierce than of any other place; the fiery rays of a torrid climate imparting a feverish heat, which animates them with an invincible courage. The smell of the Lion is not so perfect as in many other animals, but his shod of roaring supplies this defect; for, according to all report, when he roars, he puts his mouth to the ground, thus the sound is universally diffused, and not coming stronger from one place than another, the terrified animals, in their haste to escape, frequently run to the very spot they mean most to avoid; which, by a kind of bounding, he quickly secures, striking it with great force with his paw: he sometimes invades the flocks, and, with ease, will carry off a tolerable sized ox; he frequently lies couchant, as expressed by Shakespeare, "with cat-like watch," and springs upon his prey by surprise; which, if he chances to miss, in a kind of shame-

faced manner, he measures back the distance, step by step, as if to see in which he erred ——too much or too little. This animal will sustain hunger for many days, but requires a more frequent supply of drink, which it laps like a cat, and at every opportunity.—Mr. Buffon observes the courage of the Lion diminishes in proportion as his abode approaches to an inhabited country; his consciousness of man's superiority and enmity, awakens him to fear and caution; and the stately Lion, the despoiler of thousands, is frequently levelled with the dust by their address:—Three or four men usually go to the attack on horseback; these, if the Lion discovers at a distance, he takes to his heels as fast as he can; if at a small distance only, he then walks off, but in a slow and surly manner, without hurry, as if above shewing fear; as the hunters approach, he slackens his pace, eying his pursuers askant; finally making a full stop, he turns round to face them, gives himself a shake, and roars with a short sharp tone—this is now the time for the attack; he who is most advantageously situated fires his gun, and gallops off; the Lion immediately pursuing, another then fires, and by thus relieving each other, and repeating their shot, they rarely fail to reduce the tyrant of the forest. He is sometimes taken in pitfalls, but more generally when a cub, during the absence of the dam:—they may be rendered tame and docile, though, at times, subject to resume their native fierceness: their generosity and mercy is not less conspicuous than their courage; Pliny reports that " they spare the prostrate, and, when hungry, seize first on men rather than women, and never on infants, unless pressed by great hunger."

We shall close this account by referring to two remarkable instances of the memory and generosity of this noble animal which are recorded in the Guardian, No. 139, and No. 146.

THE want of that majestic and graceful ornament the mane, is the principal difference between the male and female Lion; thus unadorned, the female Lion appears scarcely to belong to the same species with the male; this dissimilarity of figure influenced us to present a drawing of the Lioness; the variations will be best understood by comparing their two figures.

The female Lion is, in general, smaller than the male; but, having the same habits, and being actuated by like propensities, we shall take this opportunity of enlarging the description already given of this noble animal. Our choice of information will, we hope, be approved when we assert, it was collected in the midst of their haunts, surrounded by their presence, and being corroborated by many witnesses, appears to have manifest advantage over the generality of reports, which usually have passed through several relators. To the diligent researches of Mr. Sparman we, therefore, acknowledge our obligations for the following particulars.

The natural instinctive dread, and strong perception in animals in general, of a Lion near at hand, is thus related: We could plainly perceive by our animals (viz. horses and oxen) when the Lions, whether they roared or not, were reconnoitring us at a small distance; the hounds then did not dare to bark in the least, but crept quite close to the Hottentots, and our horses and oxen sighed deeply, frequently hanging back, and pulling slowly, with all their might, at the straps with which they were tied up to the waggon; they also laid themselves down and stood up alternately, appearing as if they did not know what to do with themselves; and, indeed, I may say, just as if they were in the agonies of death. One of our oxen, on this and every similar occasion, appeared particularly disquieted and restless, and besides, made a remarkable noise inwardly; the same was the case with the stallion in his particular way. At night, between 10 and 11 o'clock, we heard the roaring of a Lion, and, though it roared only twice, the animals were restless the whole night through. The bounty of Providence claims our admiration in this instance, which having sent such a tyrant as the Lion among the brute creation, has likewise taught them to discern and distinguish it at a distance with trembling and horror.

The following trait is curious, and, we believe, new: That the prey of the Lion, if of the brute creation, is immediately dispatched, but if of the human species, although provoked, he contents himself with merely wounding, or, at least, waits some time before he gives the fatal blow. A farmer, who had the misfortune to see a Lion seize two of his oxen which he had just taken out of his waggon, told me, they instantly fell dead, though, upon examining, it appeared, their backs only had been broken: on the other hand, the conversation ran every where in this part of the country upon one *Bota*, a farmer and captain of militia, who had lain some time under a Lion, and received several bruises from the beast, and was a good deal bitten in one arm; but, upon the whole, had, in a manner, his life given him by this noble animal.

The strength of the Lion is unquestionably very great, yet, to strength he is sometimes obliged to add cunning;—to attack the Buffalo, he steals upon his retreat, and, by surprise, seizes the animal by the nostril with both paws, which he continues to press close together, till at length, the creature is wearied, strangled, and dies. In running they (two which we chafed) had a kind of a sidling pace, like that of a dog, accompanied now and then with a flight bound, their necks somewhat raised, they looked askant at us: they appeared to be considerably higher and longer than our horses, which were of the size of common galloways.

THE ROYAL TIGER.

Drawn from Life Engraved & Publ.d by The Cattons Jun.r who he is now. Nº 7 the Three Turnstile Court Prel. Nov.r 1794.

The ROYAL TIGER.

THE observation of the inferiority of works of art, compared to the productions of Nature, is particularly applicable to the skin of the Tiger, the beauty and splendour of which surpasses that of most other quadrupeds; and what the Peacock is among birds, in the opinion of the Ancients, the Tiger is among beasts. The colours which adorn the skin of the Tiger are, a bright yellow chesnut on the back, legs, and about half way down the sides, which are beautifully enlivened by black perpendicular stripes; the feet and tail are also marked with the same coloured stripes, but horizontal: the belly, chest, and cheeks are white, and also marked; is classed among the Cat Tribe.

Of Tigers there are several kinds of various sizes, but which still retain the same destructive qualities, and possess the same desire for carnage. The one here shewn is the largest of the species, and is called the Royal, or Bengal Tiger; which, Mr. Buffon says, is the only one deserving the name of a Tiger. "It is a terrible animal, and more to "be dreaded than the Lion; its ferocity is comparable to nothing. Let "us judge of its power by its stature; it generally stands four or five "feet high, and is nine, ten, and even thirteen, or fourteen feet long, "exclusive of the tail."

The one from which this drawing was made, as near as we could measure, was nine feet long, and about seven or eight feet from the rump to the nose, and about four or four and a half feet high.

Tigers, though slender in their make, possess great strength of body; the report of travellers in this particular, if not corroborated by numbers, could scarcely be credited. The Buffaloes of India are very large, yet a Tiger has been known to carry one on his back with such alacrity, that his speed was scarcely impeded; the weight of an inferior animal, or a man, will nothing obstruct his flight.

The Tiger, like all of the Cat Tribe (the Lion excepted), seizes his prey by surprise; lying in ambush, they wait with silent patience for an opportunity to spring on the destined victim, which he will bound upon at the distance of twenty or more feet; and if he chance to miss, does not persevere openly in the attack, but in a cowardly manner skulks about, and seeks another opportunity for effecting that by surprise, which a dastardly temper denies to his strength; yet, when urged by necessity, the Tiger shews either great courage or despair: In combating the Lion, he defends himself valiantly, and frequently with victory. The Tiger, if undisturbed, always first sucks the blood of his prey, rooting his head even into the cavity of the body; happiness appears at the highest when he drinks in the gore of the vanquished.

The rapacity of this animal engages him sometimes in conflict with the Crocodile, who, when the Tiger approaches the water to drink (which a constant thirst compels him frequently to), readily puts up his head in hopes of a prize; then the Tiger immediately strikes his claws into the eyes of the Crocodile; this unwelcome salute is declined by sinking in the water, the Tiger still retaining his hold. Thus circumstanced, the issue is doubtful, either combatant striving not less for victory than for life; the agonizing pain of the one, and the irritated fury of the other, producing a conflict truly savage: in which death is often the portion of both.

Instances have been known of the slightest occurrence, or noise, repelling or disappointing the attack of the Tiger. The Royal Tiger here shewn was scarcely known to the Ancients, and are rare in the East Indies, which may be considered as their native soil.

THE HUNTING LEOPARD.

The HUNTING LEO'PARD.

THIS animal is one of the many which belong to the Cat tribe, and posseses the same insidious disposition, and in proportion to its bulk, the same rapacity of temper, and desire for carriage.

The neck and belly of the Hunting Leopard are yellow, or a dirty white; the back and feet, a tawney brown, beautifully marked with black spots; the head is small and round; the ears short, as though cropped, which gives a very smart and lively appearance to the creature; is about the size of a large dog; the body rather long and narrow; the legs and tail rather long, particularly the latter—inhabits the torrid climates, as in general this tribe do, where their rapacity and unfatiated appetite secure to themselves whole tracts of country, the beauties of which man is obliged to give up to these savage posessors.

The larger Leopards may be considered of an untameable disposition, which is rarely or never conquered; severity will not reform, nor will kind usage soften; and while the Lion, the Bear, &c. may be brought to familiarity, the Leopard or Panther must ever be kept at a distance. The smaller animals of this species are, however, frequently tamed, and rendered obedient to the will of its governor; we have had occasion already to notice this in the Persian Lynx, and the Hunting Leopard may be added as another instance.

In India it is tamed and trained for the chace of Antelopes, carried in a kind of small waggon, chained and hood-winked, till it approaches the herd; when first unchained, does not immediately make its attempt, but winds along the ground, stopping and concealing itself till it gets a proper advantage, then darting on the animal with surprising swiftness, overtakes them by the rapidity of its bound; but if it does not succeed in its first efforts, consisting of five or six amazing leaps, misses its prey, losing its breath, and finding itself unequal in speed, stands still, gives up the point for that time, and readily returns to its master. The height, as it sits, is about three feet.

TIGER CAT

The TIGER CAT.

THIS animal, in the general outline, resembles the common domestic cat; its colour is beautiful, its shape more slender, and size considerably larger, generally between two and three feet from the nose to the base of the tail—the fur of a bright cinnamon colour, is beautifully variegated with dark stripes and spots, the former along the back and flanks, the latter mark the breast and legs; the tail, which is rather short, is also streaked; the head resembles that of the common cat, the ears are short. The Tiger Cat possesses great activity and vigilance; is indefatigable in pursuit, and bold in attacking, but fearful of man; preys on sheep, calves, and various kinds of game. Hernandez, in his History of Mexico, reports of the cunning and craft of this animal, that it will sometimes lay itself out along the branch of a tree, as if dead; thus exciting the curiosity of the monkey, whose approach is quickly followed by death. This animal is a native of America, particularly the southern parts; is met with in great plenty about the Cape of Good Hope; is frequently called the Mountain Cat, and by Mr. Buffon the Ocelot. They inhabit both the mountains and woods.

The CIVET CAT,

IS claffed by Naturalifts among the Weafels, on account of its long body, and proportionate fhort legs; and alfo, becaufe the claws are not fecured or covered with a fheath when drawn in.

The fhape of this animal is pleafant and agreeable; the colour a brownifh grey, variegated on the body and legs, with darkifh fplotches; the nofe fharp, and black at the tip; the ears rather large and round; the tail nearly as long as the body; the length of which, from the nofe to the rump, is about 28 inches, and the height about 12 inches;—is fubjeƈt to confiderable variations both in fize and colour, which in fome is cinnamon, ftriped almoft like a Zebra.

With care, this animal will live in the more temperate climates, but muft be confidered as a native of the warmer ones; viz. the Eaft-Indies, the Philippine Ifles, Madagafcar, and the Brafils; where it produces wild in great abundance; inhabits the woods, feeds upon birds and poultry; in cafes of neceffity, eats roots and herbs; poffeffes confiderable boldnefs, and is not eafily tamed, or ever familiarized.

Of this animal great gain has been made from the perfume which it produces in a glandular pouch or bag, which fprings under the anus, and hangs between the legs: this fecretion, rated as an article of com-

merce, has procured an attention to be given this animal, which, but for gain, it would never have experienced. Civet, as a perfume, fome years back, was in high eftimation; many animals were kept, and fed with attentive hands, in order to increafe and fecure this exfudation of nature. In the upper external part of the pouch is an orifice, which opens into the bag where the Civet is depofited (in appearance like foft pomatum); and fuch as are kept for profit, are confined in a long wooden box, little bigger than the animal, and the receptacle of the perfume is fearched, or rather fcraped, with a fpatula or fpoon twice or thrice every week, and the produce carefully preferved; which yields a good profit.

The Dutch bring the greateft quantities of this valuable perfume to market, and theirs is generally efteemed the beft, and moft free from mixtures, which are added to increafe the weight, but impoverifh the drug. It ftill continues a confiderable objeƈt of traffic in parts of India, the Levant, and the Brafils; of which latter place Mr. Buffon doubts the animal's being a native. The fumes of the Civet, when irritated, produce a ftrong odour; and any place it rubs againft ftrongly partakes of the fcent: the fkins will long retain the valuable quality.

THE PERSIAN LYNX.

N.7 in the Series Tuberden Kennel Feed Aug 4.1787.

The PERSIAN LYNX.

THIS animal belongs to the Cat Tribe, and is closely allied to the Ounce and Panther; and, notwithstanding Nature has bestowed a disposition fierce and savage, they frequently are made tame and subservient to the amusements of the chace. For the entertainment of the Grandees they are taken young, brought up tame, and early accustomed to the chace of both beasts and birds. They are carried to the place of diversion on horseback behind a rider, and when shewn the game, run at it with great speed, and, jumping on its back, insidiously creep forward, and scratch out the eyes; thus effecting by art what their strength could not achieve, they never openly attack any animal, but such as their superior force can readily overcome. One fell on a hound, which it killed and tore to pieces in a moment, notwithstanding the dog defended itself to the utmost. They surprise with great address the larger kinds of birds, such as Cranes, Pelicans, Peacocks, &c.

The Persian Lynx is called also the Lion's Provider, because the natives report they seek out and hunt down provision for the Lion; this is generally supposed to be the cause assigned to the effect; as they inhabit the same climate and countries as the Lion, it is most probable they frequently attend upon and partake of his leavings: the same report also informs us, that when they call the Lion, the voice resembles that of one person calling another.

The Turks call it Karrah-Kulak—that is, black ears—The Persians, to the same purpose, name it Siyah-gush. Mr. Buffon calls it the Caracal.

The size of the Persian Lynx is something larger than a full-grown fox, colour not quite so red, fur not so long, resembling more that of an hare; chest and belly part white, has a very long visage, and over-hanging eye-lid, which gives a very sly, insidious and sulky appearance; the ears large and thin, inside white, black on the outside, with a radiating tuft of hair on each of the same. Inhabits the warmer climates of Asia.

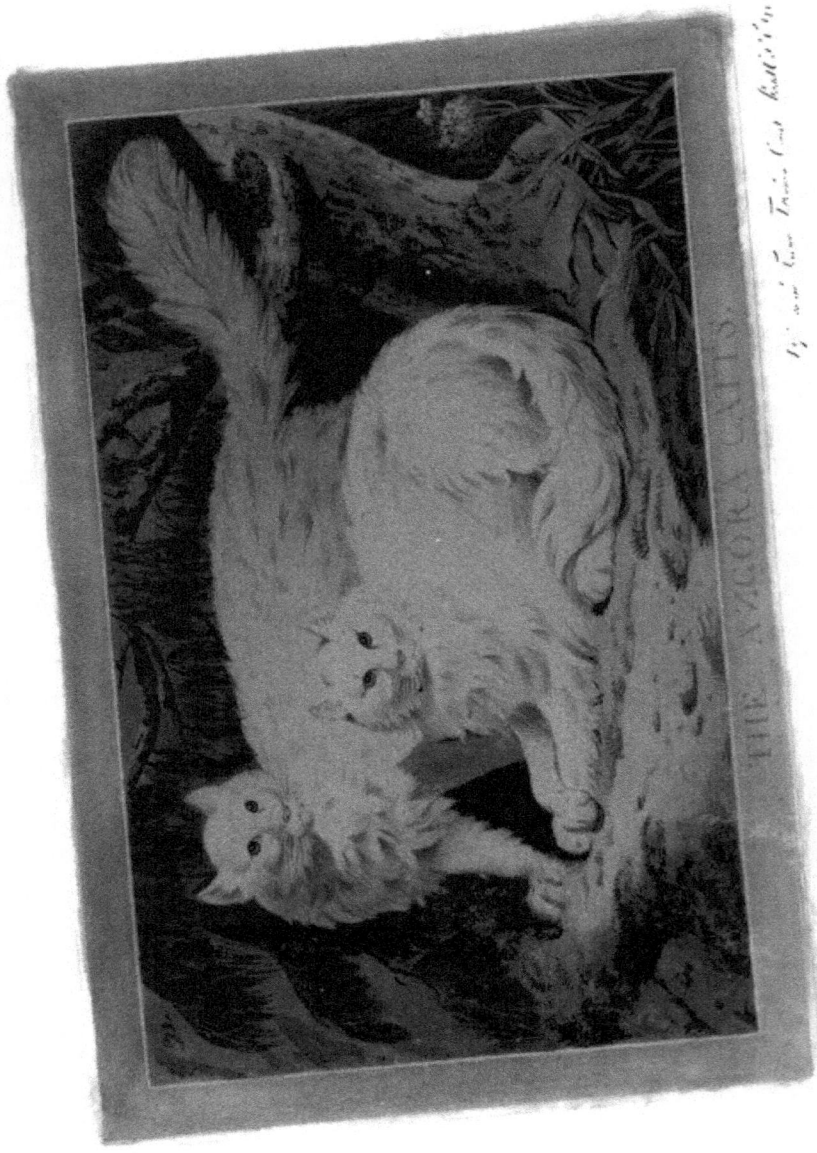

THE ANGORA CATS

The P E R S I A N   C A T.

IN our description of the Angora Goat, we had occasion to remark the influence of climate upon the hair or fur of animals; to which we shall now add Mr. Buffon's observations on that phenomenon. "In general we may observe," he says, "that of all the climates of the habitable world, those of Spain and Syria conduce the most favourably to this agreeable change of nature. Sheep, goats, dogs, cats, rabbits, &c. both in Syria and in Spain, have the finest wool, the handsomest and longest hair, with colours the most varied and agreeable; it seems as if Nature here meliorated and embellished the form of her animals. The wild cat of a harsh colour and a rough skin, when domesticated, acquires a soft skin of various colours; but in the favourable climate of Chorazin and Syria, the hair becomes longer, finer, thicker, and the colours uniformly more agreeable; the black and the red mingle to an agreeable brown, and the dark grizzle becomes a pleasant grey. In comparing the wild cat with the domestic, we discover no difference, but in the variation of colour. This beautiful appearance will not long survive the change of climate; after the first generation, they no longer attract the eye with splendour, or invite the touch with softness."

The Persian, or Angora Cats, here represented, were brought from the country whence they have their name, and were in fine health and condition when this drawing was made.

The hair was three or four inches long, of a fine milky whiteness; the eyes a lively blue; the hair on the neck longer than on the body, and the tail was bushy; for the rest they resemble our common cats, only appearing rather larger, on account of the greater length of the fur; they possessed the same habits, and were alike playful and insidious: we have every reason to believe they were entirely deaf. They are now at Mr. Brook's Menagerie, London.

THE YELLOW—BEAR.

Drawn from the Original & Published by Ch.º Catton jun.ʳ   Nᵒ 5. in the Terrace Tottenham Court Road Oct. 1ˢᵗ 1788.

The YELLOW BEAR.

O F the Bear, as well as of moſt other animals, there are ſeveral varieties, which in general are well known; the one here repreſented, we have reaſon to ſuppoſe, was never before drawn, or deſcribed; our reſearches having not yet met with any account of the Yellow Bear.

The Yellow Bear from Carolina (as the American Bears in general are) is rather ſmaller than the European Bears; it has alſo a more pleaſant and agreeable countenance; is perfectly tame and ſociable; the colour a lively bright orange, of a reddiſh caſt. The hair thick, long, and ſilky. Its other properties are the ſame as to the ſpecies in general. This drawing was made from the animal which is now in the Tower at London.

Of Bears in general we may obſerve, they are inhabitants of the colder parts of Europe, Aſia and America; and for the moſt part are not carnivorous, but feed on fruits, roots, berries, and vegetables; fiſh alſo form a conſiderable part of their food, in catching which they practiſe great dexterity. Of potatoes they are partially fond, and when once they diſcover them, it is with difficulty they are kept from having the greateſt ſhare. The paw of a Bear is as well calculated for rooting up the ground, as the ſnout of a hog. Are fond of peas, which

they gather and beat out of the huſks on ſome hard place, and carry off the ſtraw; but tread down and deſtroy more than they conſume.

In winter they are hunted for the fleſh as well as the ſkins, which make a conſiderable article in the fur trade. Among 500 Bears killed one winter near James's River in Virginia, there occurred but two females; this being the ſeaſon for their breeding, nature dictates the moſt guarded ſecluſion, left the young ones ſhould be devoured by the males; of theſe two neither was with young.

Bears, notwithſtanding their clumſy appearance, are very nimble creatures, and will climb the higheſt trees with ſurpriſing agility, and, if wounded, will defend with great fury and reſentment to attack the aggreſſor, who, without armed aſſiſtance, has a bad chance for ſafety. In defending themſelves they ſtrike with the fore feet like a cat, and ſeldom or ever uſe their mouths, but ſeizing the aſſailant with their paws, and preſſing him againſt their breaſt, almoſt inſtantly ſqueeze him to death.

Bears, after feeding on autumn's plenty, are very nice eating; the fat is particularly mild, and more may be eaten of it, without offence to the ſtomach, than of any other fat.

Drawn from the Original & Engraved & Published by Chas. Catton Jun.

No. 1 Pen at Terrace Fountain Cross Road   March 1788.

A N I M A L of the B E A R - K I N D.

THE wonderful variety of quadrupeds, and the frequent variations in each species, lead the thinking mind at once to admire the boundless productions of nature, and to confess, that man with all his acquirements of knowledge, is but very slenderly informed, and at best but superficially wise; that more knowledge and additional discoveries tend less to perfection, than to excite inquiry after a train of endless researches.

The animal we are now to describe is of a form unknown, of a species never described, with peculiarities which have never presented themselves, or else have escaped notice.

The " PETRE BEAR," (which is the name we have been directed to call it by a very able naturalist) was brought to England in August last, by Capt. Pearson, on board a ship belonging to the India Company, whose report is, that it was brought from Patna in the province of Bengal, and when it first came unto his possession was very young, nearly a cub; —here information leaves us: the manner in which it was caught, or the place and circumstances attending, were not related; we must therefore rest contented with a description of its several parts, noticing such peculiarities, as have been observed since it came to England. The animal is a female, and is now to be seen at the beast shop, Holborn-hill, London.

The Petre Bear (if the name be allowed) has five claws on the fore feet, between two and three inches long, nearly of the same thickness, and not very bulky: the hair all over the body is of a rustyish black colour, very harsh and coarse, between twelve and eighteen inches long; that on the shoulder rather longer, which it can draw forwards, or lay backwards: the form of the head, and chiefly the mouth, is the principal characteristic of the animal; this we have endeavoured to ren-

der as intelligible as possible, by adding a view of the mouth when open: from the eyes to the tip of the nose, is about 6¼ inches, which tapers off like the truncated snout of swine: the front of the mouth, when shut, is flat also, like the swine: the formation of the nostrils differs from every other animal's, the natural shape being exactly as shewn in the drawing: the lips project very far (two or three inches) beyond the front teeth: the lower lip, as well as the upper one, with the nostrils, are very flexible, over which the animal has great command, placing them in any position at pleasure; has great power of suction, and will draw things into its mouth at a considerable distance: the tongue is rather small: teeth at present few: was fed on board the ship, with boiled rice and sugar; at present on bread (about a quartern loaf per day): with some difficulty was brought to eat flesh, which it does now with a relish: is very tame; appears pleased when kindly noticed, soliciting play: expresses anger with a kind of accented growl, something like barking: its paws, when laid together, it sucks with a tremulous noise, like one when shaking with pain: the feet are short; has nails on the hind legs: drinks, or rather sucks water in considerable quantity: general appearance black, face grey, stands about 2½ feet high, 5 feet long, has grown considerably since it has been in England.

Upon the whole, this may be considered as a very uncommon creature, as well for its formation, as the very extraordinary circumstance of so thick and coarse a fur to a native of so warm a country; the general appearance also is that of a Bear, whereas we know of none in those climes, nor any animal whose appearance can lead us to suggest to what breed it belongs, or by what mixture or chance it is likely to owe its birth. Time and further inquiry only can clear up these at present singular facts.

THE SWEEDISH–ELK.

Drawn from Nature & Engrav'd y Publish'd by Cha.ᵉ Catton Jun.ʳ — N.ᵒ 5...on the Terrace Tottenham Court Road Oct.ʳ 1 1788.

The ELK.

THE animal from which this drawing was made, was a native of Sweden, the dimensions and other particulars of which we shall first enumerate.

The length of the head was two feet, the body three feet, and neck one; the total height was six feet, of which the legs were three; the ears were about twelve inches long; the general colour a blackish brown brindled, some hairs being brown, some white, and others partaking of both colours; the lightest colour was on the neck, and upper parts of the body; the knees of the fore legs, and inner parts of the hind legs white, which extended but little on the brisket; the hair universally very coarse and thick; about two inches and a half long; considerably longer under the lower jaw, and along the upper part of the neck, resembling an upright mane, which was longest and fullest over the shoulders; the tail not more than two or three inches, was nearly lost among the hair of the body; the head long; the upper jaw about the nostrils very full and chumpy. We have been seen in England, few having knowledge of this animal, one belonging to his Grace the Duke of Marlborough; the other now exhibited at Mr. Parkinson's Museum, London, from which this drawing was taken.

Of the general history of the Swedish Elk, the following particulars have come to hand. In walking, the Elk carries its head nearly horizontal, and when passing the thick parts of the forest, the head is so disposed that the horns lay close upon the shoulders; thus the inconvenience of such wide extended antlers is much avoided. The horns of this animal are reported to grow to very surprising dimensions; the fossil horns frequently found in Ireland are supposed to belong to the Elk, or an animal of that species: some have been found extending fourteen feet; those to the animal here represented, were distant about four feet at the top; each horn having five tips, leads to suppose the animal was about six years old, and certainly not yet come to its full growth. The climate of Sweden in general, but of the northern parts in particular, being severely cold, nature has not only provided a thick warm fur, but has so formed the under part of the hoof, with a very sharp edge, that this animal with perfect safety can pass over the smoothest ice. The race is not always to the swift, nor the battle to the strong, else this animal would be secure from harm; the bear and the wolf hunt it with success, the shape of their feet, and the lightness of their weight, enabling them to travel the frozen snow with safety; while the flight of the intimidated Elk is impeded by frequently falling through the frozen crust, which not only flops its progress, but also wounds its heels; thus suffering from pain which is increased by repeated disasters, it happens ere long the enemy seizes it by the throat, and the largest European animal is conquered by one, not a quarter of its size, or weight. This animal, when defending itself, uses not only its horns, but rising on the hind legs endeavours to plunge the fore ones into the body of its adversary.

It feeds upon herbage, prefers the aspen tree, resides on uninhabited islands, in large rivers, and on the banks of great lakes. The report of its speed is, that it will travel 300 miles a day; the flesh rather brown and coarse, but good tasted.

The A N T E L O P E

FORMS a diſtinct claſs of animals, partaking many of the charac-ters of the goat and the deer, yet diſtinct from either.

The Antelope, or Gazelle, has hollow, truncated and permanent horns, which is not ſo in the deer; declines the graſſy paſture, and browſes on ſhrubs, which imparts to the fleſh a delicate and agreeable flavour; on the other hand, the ſize and delicate form agree with the deer; the colour and nature of the hair alſo is the ſame, but the form of the horns differs very much, being of one ſtem or ſtalk only (without any branch-ings out), though of various and very different inflexions, which mark the ſeveral ſpecies of this tribe. The horns are annulated or girt round with rings at various intervals, at the ſame time they are longitudinally depreſſed or flattened from top to bottom; theſe particu-lars are common to all of the numerous race of Antelopes; of which Mr. Buffon has enumerated twelve; Mr. Pennant doubled that number, and perhaps there may yet be many undeſcribed varieties, if variations ſo ſlight, and marks ſo ſmall, conſtitute a new kind. The influence of ſoil, country and climate, has moſt likely produced this great variety, and where it may end is too uncertain to ſay. In general, they inhabit the hotter climates of Aſia and Africa (three or four kinds only ex-cepted), and go in companies of ſix or eight, or elſe herd together in vaſt quantities.

Mr. Sparman mentions a lucky eſcape he and his travelling equipage experienced, in being about a quarter of a mile on one ſide a herd, not leſs than ten thouſand, which took their courſe over the plain where he lay encamped, and, but for this fortunate ſpace, he and his compa-nions would have been trodden to death.

The eyes of this animal are the moſt beautiful and meek which Nature has formed. The eye of the Antelope is a metaphor in general uſe among the Eaſtern poets; and the gallantry of a lover, in thoſe countries, can go no higher, than in comparing the eyes of his miſtreſs to thoſe of the Antelope.

The form alſo is very elegant, and, with their ſwiftneſs, is noticed by the Sacred Writers. Their activity is wonderful, and their ſpeed exceſſive; the fleeteſt dogs are left far, very far behind; and when hunted, the aid of the hawk is neceſſary, which, by faſtening on the neck and cheek, either mortally wound, or ſo much impede their flight, that an opportunity is given for the horſemen and dogs to come up and ſecure the game. The Lynx or Panther is often employed in this chace, for the amuſement of the Great; theſe animals ſucceed more by craft than ſpeed; creeping ſlyly forward in a winding courſe, at an unwary moment they ſpring upon the thoughtleſs animal, mortally wound, and ſuck the blood.

Antelopes are very timid; ſome dwell upon the plains, others, and moſt generally, among the hilly countries; at the leaſt or moſt diſtant alarm betake themſelves to flight, and ſeek for ſafety in dangerous and inacceſſible parts of the rocks; where, it is reported, the Antelope will ſtand upon a pinnacle or point, no broader than the ſpace occupied by the four feet drawn cloſe together.

The one here delineated is the common or brown Antelope, the horns of which are about eighteen inches long, and fifteen diſtant at the tips; ſtands about three and a half feet high; colour a bright brown; cheſt, belly, &c. white.

THE WHITE-FOOTED ANTELOPE.

The WHITE-FOOTED ANTELOPE.

THE great variety of Antelopes, and the material difference between this and the one before given, induced us to prefent our readers with this drawing of the White-Footed Antelope. In our former account we remarked the almoft innumerable varieties of this tribe, with marks effentially different, yet partaking of the fame general character and properties.

The Antelope now before us, is marked by four white feet; the general appearance alfo varies confiderably; the colour is a dark or blackifh brown; on the cheeks are two white fpots, and on the neck there is a tuft of black hair.

Its ftrength and activity were very great; much pains and labour were beftowed before it could be brought to be the leaft fubmiffive or familiar; its height was equal to that of a common galloway, or fmall horfe; the hoofs were long, and divided far up.

Having before prefented our readers with the leading features and principal characteriftics of the Gazell or Antelope tribe, we beg to refer to a comparifon of the two drawings, for the more minute variations of thefe two animals of the fame fpecies.

THE MUSK DEER.

Drawn from Nature Engraved & Publd by Chs Catton Junr.

Nᵒ in the Inner Temple Lane Price Publd Nov 1 1788

The M U S K - D E E R.

OF the Muſk Deer much doubt and various opinions have been held, by travellers and naturaliſts, in which claſs of quadrupeds to place it; ſome contend for the Horſe, ſome the Ox, and other the Deer; the latter appears the beſt deciſion, notwithſtanding the animal wants thoſe ſtriking characteriſtics the horns; ſuffice it for us, the animal is known by the name of the Muſk-Deer. The figure here drawn, will better ſhew its form, than any deſcription.

This animal is particularly known by producing that celebrated drug the muſk, which heretofore was in high eſteem as a perfume, and at this time is conſidered as a very powerful and valuable medicine in nervous and other caſes. In the male animal only, this exudation gathers in a ſmall bag or cuſhion, between the legs, near the groin, in ſubſtance like coagulated blood, or, as ſome deſcribe it, a brown fatty matter: indeed certain information is wanting of this particular. Mr. Gmelin reports, that when the bag by being over-full becomes troubleſome, the animal expreſſes part by rubbing againſt a tree, or other convenient place; the matter dropping in ſmall parcels on the ground is ſecured, and is eſteemed by much the beſt muſk.

Others again report, that it comes to market only in the repoſitory formed by nature, which is not bigger than a ſmall egg, and that the weight of four is not more than an ounce. Conſidering its ſmallneſs, the animals muſt be in great abundance to ſupply the quantities uſed in Europe only; in the Eaſt its reputation being much greater, its uſe muſt

be more general. The male animal only producing this treaſure, muſt be hunted and killed for the bag. This drug, like every other article of value, is very liable to adulterations; ſmall pieces of lead are thruſt into the bag to increaſe the weight, and foreign matters often mixed with it, to augment the quantity.

To the mouth of the male animal there grow two teeth or tuſks from the upper jaw, rather bent; theſe are very white; to the female theſe are wanting; the latter alſo is much ſmaller than the male. The fleſh is ſtrongly impregnated with the perfume of muſk, but is neverthleſs eaten by the Tartars.

They inhabit various parts of the Eaſt Indies, China, Tartary, and Siberia, from which latter place the one here drawn was brought. The colour, a reddiſh brown, marked on each ſide the throat with a ſtripe of white, and ſeveral ſplotches of the ſame on the flanks; the attitude of the hind legs was alſo very aukward, as expreſſed in the drawing. Naturally dwell among the mountains, and feed on herbage, and the young buds of trees, particularly pines: prefer ſolitude, and avoid mankind: if purſued, flee to the higheſt ſummits, inacceſſible to men or dogs.

The ſize of the one from which this drawing was taken, was, from the noſe to the inſertion of the tail, two feet four inches; from the ground to the ſhoulder, twenty inches.

2

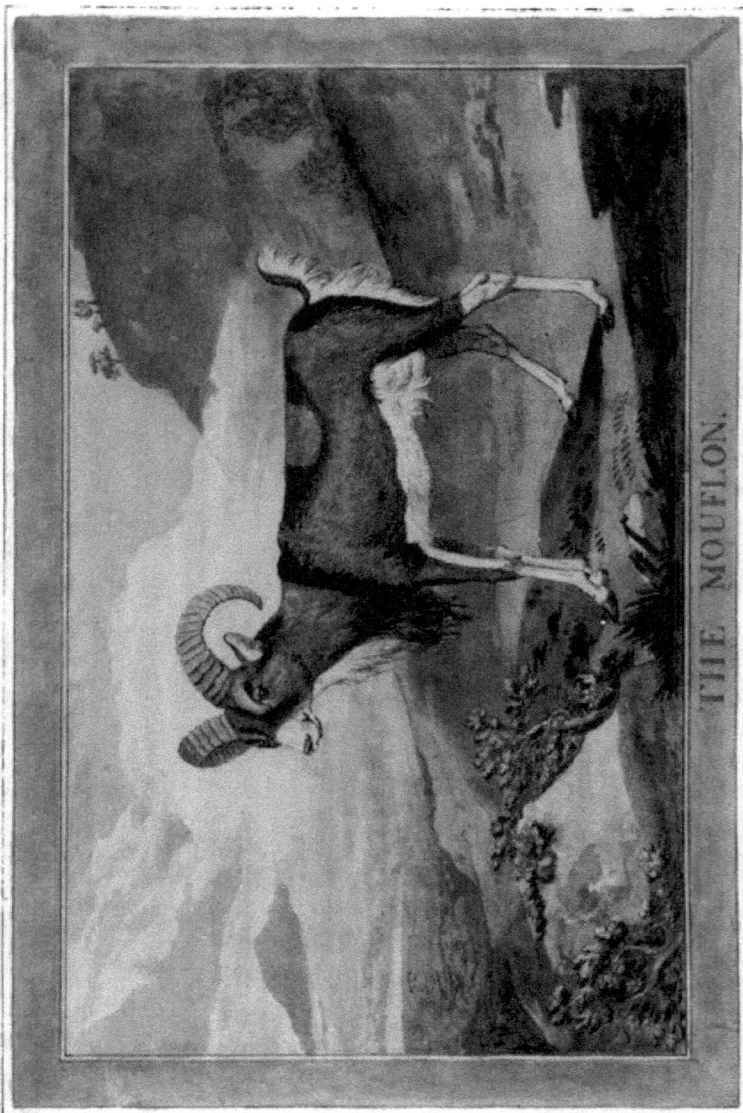

THE MOUFLON.

Drawn from Nature Engraved & Publish'd by The Catton Jun.r

Nº. Van.r Senr Tottenham Court Road March 1 1790.

The MOUFLON.

NATURE with a wife and provident care gave to every animal originally either force to repel, speed to escape, or cunning to evade its more formidable enemies, as a means of preserving the continuance of her various productions. Of useful animals the weak were first reduced to the services of man, smaller animals more readily adopting the influence of education. Thus the Sheep and Goat were first brought to usefulness, before the robust Ox, or vigorous Horse ; this may be considered as one principal cause of the endless variety of such as have been long and particularly attended to by man ; and the power of continued education, joined to the influence of climate and soil, make it difficult to say, which are the true characteristic marks, or which the original of many species.

The Mouflon is considered as the Sheep in a state of nature, by Mr. Buffon, with a temper not broken by servitude, a constitution not softened by inactivity and luxury. With a vigorous mind it defends itself against the attacks of larger animals, and, aided by a robust body, frequently overcomes formidable enemies.

The abode of the Mouflon is in rocky countries, where they bound from rock to rock, or climb the apparent inaccessible precipices with that address and ease, which characterize the Goat and Deer tribes, and which sets pursuit at defiance. The horns of the Mouflon are very broad at the base, are firmly fixed upon the skull, inclining backwards with a considerable curvature, the distance increasing to the extremities ; the horns, of a light brown or yellow, are girt with many annuli or rings ; in the male they frequently grow to very large dimensions, and weigh sometimes thirty pounds ; and, when broken off, as in defence, or by other accident, often serve as a nest, or retreat, for various small quadrupeds, such as young foxes, &c.

There is a beautiful form in this animal, which approaches very close to the Deer ; indeed by some it is esteemed the same as the old or original Welch Deer ; however, the shape of the head, and the truncated horns, which are never shed, rather mark it of the Sheep tribe. The major part of the body is of a fine brown coloured hair, which on the lower part of the neck and chest grows to a considerable length; the belly, legs, and outer part of the haunches are white.

Inhabits the warmer parts of Europe, such as Greece, the Isle of Cyprus, Sardinia, and Corsica ; they also are found in great numbers in the southern and mountainous parts of Siberia, a climate rather cold than temperate ; grows to the size of a young stag. The measurement of the one here drawn was three feet from the nose to the rump, and two feet and a half from the shoulder downward.

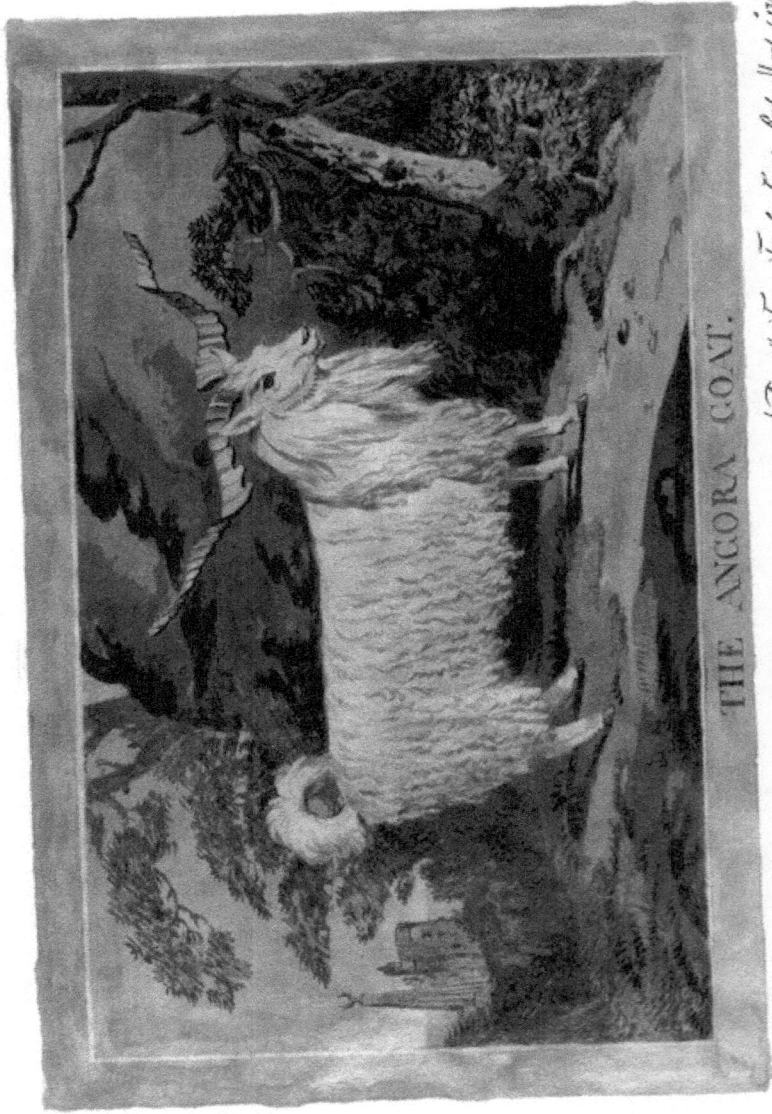

THE ANGORA GOAT.

Drawn from Life Engraved & Published by Charles Catton Jr.

Nᵗ.₁₀ in the Terrace Tottenham Court Road March 1788

The A N G O R A G O A T.

THIS animal affords the moſt ſtriking example of the influence of climate upon the clothing or fur of Goats and Sheep. When transferred from a cold or temperate country to a hotter one, the fine wool of the ſheep degenerates to a coarſe hair; and a few miles of country ſurrounding Angora in Aſiatic Turkey, produces this very extraordinary variation from the common appearance of Goats. The climate of theſe countries, and Syria in particular, impart to all the quadrupeds, a ſplendour and fineneſs of fur, which is not to be equalled in any other part of the world.

The horns of the Natolian or Angora Goat are of great length in the male, and lie in an horizontal poſition, twiſted like a cork-ſcrew; to the female the horns are very different, the form nearly a complete circle round the ear.

The hair of the Angora Goat exceeds in beauty that of every other animal; the ſilvery locks formed by Nature's careleſs, though graceful hand, flow in ringlets of about nine inches long, with a ſilky fineneſs, and reſplendent whiteneſs.

The hair of this Goat, for its beauty and fineneſs, is bought by all nations, and is the baſis of our beſt camblet fluffs; it is alſo wrought into the uſeful article called mohair. Mr. Pennant reports, the hair is imported in the form of thread, as the Turks will not allow it to be exported raw, as the ſpinning gives employ to a multitude of poor. Much pains is taken by the owner of the flocks, in keeping the animals clean, and often combing them. Of late, ladies muffs in England have been frequently made of the hair of this animal, the delicacy and graceful form of which produces a pleaſing and agreeable effect.

The one this drawing was made from was conſiderably larger than our Goats; the meaſurements were three feet ſix inches from the noſe to the rump, and two feet two inches from the ſhoulder downward.

a

The HYAENA.

IS in length about four feet, in form between the wolf and hog; the head refembling the wolf's, but rather fhorter, and black from the nofe to juft above the eyes; the colour a greyifh brown, marked down the body and legs with darker ftripes, inclining to black; ftands remarkably high before, and low behind, being in front about three feet and a half, and behind about two feet and a half high; is very lank or thin in the body, with a ridge of briftles like a hog, all along the fpine, and a brufh tail.

The Hyaena is a remarkably unfociable and folitary animal, dwelling in the holes and chafms of rocks, or in dens which it has formed in the earth; it is found in high, mountainous countries, in the mountains of Caucafus, the Atlantic chain, parts of Syria, Perfia, and Barbary, the moft dreary and fterile parts of the Torrid Zone, of which, fays Mr. Buffon, it is a native.

This animal poffeffes great ftrength of body, and fiercenefs of courage; it will refift the attacks of the Lion—(Kæmpfer relates that he faw one which had put two Lions to flight)—he never declines a combat with the fierceft of the foreft, and feldom fails to conquer—poffeffing great cruelty and fiercenefs, he is generally reckoned untamable. Mr. Buffon mentions to have feen one, and Mr. Pennant another in a domeftic ftate; the latter Gentleman thinks, if taken very young, they may be reclaimed by good ufage, but obferves, they are commonly kept in a ftate of ill humour by the provocations of their mafter.

Hunger feems ever to torment them with an infatiable voracity; they greedily devour whatever comes in their way, and will even root up the contents of the grave, and with a keen appetite confume the putrid corps.

The ancients had very ftrange notions of this animal, believing that it changed fex every year; this opinion, no doubt, originated from the tranfverfe orifice which is between the tail and the anus; they alfo believed that a ftone was found in the eye which imparted the power of prophecy. Mr. Shaw, in his travels into Barbary, obferves, the Arabs always bury the head of this animal, whenever they kill one, left it fhould be applied to purpofes of magic; it was alfo believed, the Hyaena had the property of imitating the human voice, and thereby feduce unwary travellers within its power; this the experience of modern times does not confirm, yet a gentleman told us, he once heard one make a noife refembling the laughing of a man, when the keeper had juft given him fome provifion, which he made as if going to take away—this was, probably, no more than a note of difpleafure, which every animal poffeffes, fuch as the growl of a cat with a moufe when any one approaches.—To conclude, no words can convey an adequate idea of the voracious and fierce afpect of the Hyaena, or of his uncouth, ill-formed fhape.

THE WOLF.

The W O L F.

THIS animal bears a strong resemblance to the dog; it has a long head, a sharp nose, pointed and erect ears, a bushy tail, hanging more between his legs than a dog's generally does, long legs, and large sharp teeth.

This animal abounds in every quarter of the earth, it inhabits large extensive forests, frequently making incursions on the campaign country in search of prey; its food is such animals as its superior strength can vanquish; lambs, calves, and sheep, which he carries off with ease and agility; the muscles of his neck are so strong, that in running off with a large sheep, it shall not trail on the ground. This animal is remarked for great cruelty; whenever he gets among a flock, he destroys with a wanton pleasure all that come in his way, before he secures his particular prey. It is reported, if a Wolf once tastes human flesh, he ever after prefers it, and will seek it with great avidity and resolution, and sometimes attack the shepherd in preference to the flock. Troops of this destructive animal have been known to attend the march of an army, and with eager appetite devour the slain in battle, or such as were superficially interred:—in general, this animal pursues his prey alone, but when any great business is to be achieved, such as to attack an ox, deer, or other large animal, they will assemble in numbers, and commit their depredations in consort; but when the business is finished, the assembly breaks up, and each retires to his hiding place. The scent of this animal is remarkably good, as he will wind his game at many miles distance. To hunt the wolf is esteemed good sport; dogs of great courage and speed are requisite, as they sustain a long chase. When pressed with hunger, which he long sustains, if well supplied with water, this animal exercises great courage and cunning: The destructive properties of this animal make him every where obnoxious; two or three will cause a long and constant alarm to a whole country; enmity therefore between him and man is perpetual. England abounded with this scourge of rural life, till king Edgar respited the punishment of a criminal for a certain number of wolves' tongues, and changed the tax of gold and silver paid by the Welch into an annual tribute of three-hundred wolves' heads. King Edward I. entirely to extirpate them, appointed a superintendant to assist in their destruction; which was not effected in Scotland till about the middle of the last century. The sagacity of the wolf is so active, that, Scheffer reports, he will not attempt an animal, if a cord, or halter is round its neck, suspecting a trap. The wolf is about the size of a large dog, or three feet and a half long, and two feet and a half high; colour a light brown, with a mixture of black and grey; has a penetrating eye of a grey colour, which adds a wild appearance to the ferocity of his countenance.

THE OTAHEITE-DOG

The OTAHEITE DOG.

THE modern difcoveries in the South Seas, which have been fo accurately reported by that celebrated navigator, Captain Cook, have, with a new track for fpeculation, opened a new field for obfervation; and the productions of thefe ifles have been fought after with as much curiofity and impatience by fome, as a buried vafe from Herculaneum, or the remains of the Capitol by others. Under this perfuafion of the pleafure of novelty, we prefent our readers with a portrait of an Otaheitean Dog: which, as an animal of infinite importance, as an article of luxury with thefe people, may deferve fome notice from us.

The Otaheite Dog is about the fize of our large fpaniels, and nearly refembles them in appearance; the head is rather longer and deeper, or flatter perpendicularly; the ears are erect like the wolf's; the limbs appear rather larger; the colour, for the moft part, white, with lively brown fpots or blotches.

The eftimation of things in general depend much upon their abundance or fcarcity; and with an Otaheitean, whofe quadrupeds are but two, it will not excite much furprife, that thefe are attended to with fome anxiety, particularly fo, when the pleafure of the palate is concerned: this is an influence to which the moft favage nations pay refpect.

To give fome idea of the importance of a dog in the South Seas, we fhall prefent the report of Captain Cook on this particular, as given in his firft voyage : " We all agreed that a South Sea Dog was little infe-" rior to an Englifh lamb; this excellence is probably owing to their " being kept up, and fed wholly upon vegetables." Thus much as

evidence of their delicacy; the manner of cooking this dainty fhall clofe our account. Of one prefented to the Captain, we read, " the " dog was killed by holding the hands clofe over his mouth and nofe; " an operation which continued about a quarter of an hour : while this " was doing, a hole was made in the ground about a foot deep, in which " a fire was kindled, and fome fmall ftones placed in layers, alternately " with the wood, to heat; the dog was then finged by holding him over " the fire, and, by fcraping him with a fhell, the hair taken off as clean " as if he had been fcalded in hot water; he was then cut up with the " fame inftrument, and his entrails, being taken out, were fent to the " fea, where, being carefully wafhed, they were put into cocoa-nut " fhells, with what blood had come from the body. When the hole " was fufficiently heated, the fire was taken out, and fome of the ftones, " which were not fo hot as to difcolour any thing that they touched, " being placed at the bottom, were covered with green leaves; the dog, " with the entrails, was then placed upon the leaves, and other leaves " being placed upon them, the whole was covered with the reft of the " hot ftones, and the mouth of the hole clofe ftopped with mould; in " fomewhat lefs than four hours it was again opened, and the dog was " taken out excellently baked, and we all agreed that he made a very " good difh.

" The dogs which are here bred to be eaten tafte no animal food, " but are kept wholly upon bread-fruit, cocoa-nuts, yams, and other " vegetables of the like kind."

The G R E A T B A B O O N.

B EFORE we enter upon the defcription of this creature, it may
not be amifs juft to premife, that Naturalifts have divided the
Monkey tribe into three claffes; viz. Apes, Baboons, and Monkeys.

Baboons are charaĉterifed by having fhort tails, canine teeth, and the
lower part of the face prominent or truncated, like the fnout of fwine:
In general, they, with the reft of the Monkey kind, go upon all fours;
though fometimes, and the Ape more frequently than the Baboon,
or Monkey, go erect: the pofition defigned by Nature was certainly
prone, and when in their wild flate, and uneducated, this is evidently
their manner of walking; elfe would the fore feet or paws have more
fofnefs of fkin, and lefs callofity than they really have; in this particular the fore feet differing nothing from the hinder ones, which muft
neceffarily be conftantly trodden on: the fingers of the feet are armed
with long fharp claws or nails.

Of Baboons it may, in general, be faid, they are fierce and ferocious,
of great bodily ftrength, perverfe habits, and very libidinous: extraordinary ftories are related of their partiality to, and anxiety for, the
females of the human race.

The one here reprefented is called the Great Baboon, and was in
height about 5 feet; from the head to the rump was about 2 feet 9
inches; the length of the arms about 1 foot 9 inches; general colour a
greenifh black, mixed with brown; acrofs the hams a patch of purple;
a bright vermilion ftripe up the nofe to the eyes; the cheeks a dark
violet blue, with horizontal ftripes of white; the eyes fmall, and approach very clofe to each other; long hair on the head and fhoulders in
confiderable quantity, a little brighter than the reft of the body. The
inner part of the cheeks, that is, the fpace between the cheek and the
teeth, is very capacious, and ferves this fpecies of animal as a pouch,
in which to ftow provifions, &c. Feeds on roots, fruits and herbs, naturally is not carnivorous, eagerly fond of fpirituous liquors and wine—
inhabits the woods of the hotter parts of Africa.

The LION MONKEY.

THE Monkey tribe is divided into three classes—Apes, Baboons, and Monkeys.

Monkeys are characterised by having a long tail; and are again subdivided, by Mr. Buffon, into two classes; distinguished by the use they make of their tail in many of the frisky manœuvres and tricks this species of animal is so famous for. To the one the tale is called prehensile, and serves almost the purpose of another hand; for this they readily twist round the branch of a tree, and by it suspend themselves hanging in the air, head downwards; it also secures them in their seat, while the feet are otherwise employed. The other kind do not enjoy this useful property of the tail; which, Mr. Buffon has observed, belongs to none of the Monkey tribe of the old Continent: on the other hand, those of the old world possess a cavity or pouch on each side the jaw, which serves as a store-room for provison, and which those of America do not enjoy. Thus the diminished activity in the usefulness of the tail is balanced, by an opportunity of laying in store that, which else, might not at all times be so readily acquired.

The Monkey here represented is called the Lion or Silken Monkey; the colour and appearance of the hair about the shoulders resembling that of the Lion; the hair over the whole of the body is long and very fine, with a most beautiful and silky appearance; the tail is long. This animal, when sitting, did not exceed 10 inches in height.

It is the practice of the Lion Monkey to take up his abode in a large melon, or gourd, which having previously excavated, and lined with soft cotton, forms a comfortable habitation.

It is a Native of South America, particularly Guiana, and the Brasils; is rather delicate, but gentle and frolicksome.

THE CHILD of the SUN

Drawn from Life Engraved & Published by Chas Catton Jun.r N.Do the Terrace Tottenham Court Road Dec.r 1794

THIS animal is hitherto a non-defcript, belonging to the clafs of Baboons.

The only one we have knowledge of (the one from which this drawing was taken) was exhibited in England about four years back, and was reported to have been brought from South America. The head, in proportion to the other parts, was remarkably large; the contour or fhape of the face alfo rather fingular, for one of the Monkey Tribe; the fkin of the face fmooth, of a fallow complexion. The height of this animal, when erect, was five feet, and to its great ftature was joined great ftrength of body; the hair from the fhoulders all over the body was very long, but not coarfe; the colour a light fpeckled grey, refembling a Guinea fowl.

The cunning and fubtlety of the Monkey was very apparent, and in all its actions it clofely imitated the human: the hams or buttocks were bare, and of a bright vermilion colour. The fkin of this animal, when dead, was depofited in the Leverean Mufeum, where it is now to be feen.

Having already noticed the characteriftics and leading properties of Baboons, we fhall here add fome further general account of them.

"The Baboon is a gregarious animal, herds together in great numbers, and mutually unite their ftrength to repel danger, and procure fubfiftence. The œconomy of Baboons, in general, is well regulated, and thofe of the Cape of Good Hope, we are informed, obferve a fort of natural difcipline, and go about whatever they undertake with furprifing fkill and regularity; not being carnivorous, an herd of hundreds confume great quantities of fruits, &c. Grapes, apples, and garden fruits in general they are particularly fond of, and when they fet about robbing an orchard or vineyard, centinels are always placed to give early notice of the approach of danger; thefe neceffary precautions taken, the plan of operation is as follows: part enter the inclofure, pluck the fruit, and chuck it to their neareft fellow without the fence; a regular line of communication being formed from the fcene of operation to the place of retreat, the plunder is pitched from one to another all along the line, till it is fafely depofited at head quarters, which ufually is in fome mountain. During thefe manœuvres the centinels keep a clofe look out, and if danger approaches, a loud cry is the fignal for retreat; this is done in a very quick, but not improvident manner, as each one loads himfelf in the mouth, hands, and under the arms; if clofely purfued, the latter parcel is firft dropped, next that in the hands, and laft of all, if very much preffed, that in the mouth. They commit their depredations with fuch boldnefs and addrefs, that the natives, to protect their property, are fubject to frequent watchings, and neverthelefs fuffer great damage." Kolben's Cape of Good Hope.

THE MAUCAUCO.

The M A U C A U C O.

THIS animal claffes in the rear of the Monkey tribe, and ferves as one, to connect the gradation from that fpecies to the complete quadruped, and belongs particularly to the clafs called Maki. The hands are ufed the fame in this fpecies as in the complete Monkey, ferving for every purpofe of feeding, climbing, and playing.

The nofe of the Maucauco is long and flender, black on the tip; the eyes very large and fine, furrounded with a circle of black hair; the ears large and upright; the arms, feet, and toes, or claws, like the Monkey's; the tail beautifully marked with alternate black and white rings, and is confiderably longer than the body, which is flender, of a pale brown, or afh colour, fomewhat darker along the fpine; neck and belly white. The hair is beautifully fine and foft, and ftands erect, nearly as the pile of velvet.

The Maucauco is a native of Madagafcar, and the neighbouring ifles; is very good-natured and frolickfome; poffeffing all the motion and alacrity of the Monkey, without its malice or mifchief, and is very cleanly; has a weak cry, is eafily tamed; in a wild ftate go in troops of thirty or forty.

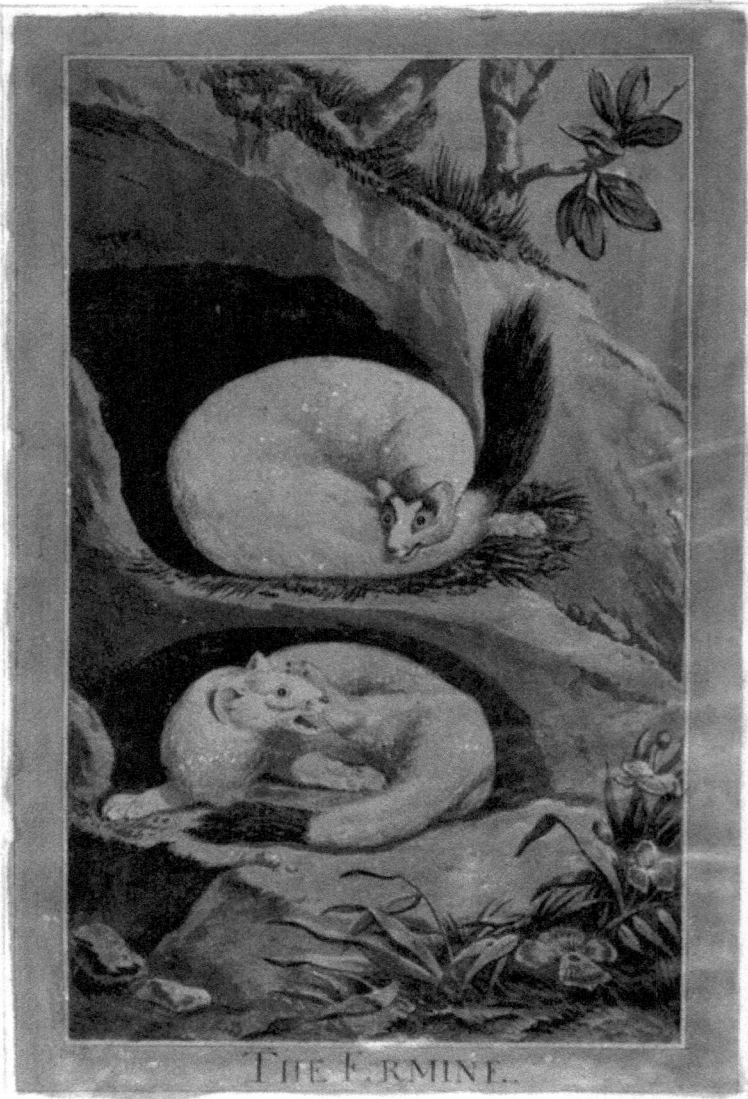

THE ERMINE.

The E R M I N E.

T H I S little animal is pretty well known on account of the high estimation in which the skin is held. The fur of Ermine is an article of confiderable commerce in the more northern countries; its delicate whitenefs and closenefs being furpaffed by none.

The Ermine is about nine inches long, exclufive of the tail, which is between five and fix inches more; the tale is black, and in the male part of the forehead is dark brown.

The animal called in England the Stoat is the fame as in more northerly climates is called the Ermine. It is a curious phenomenon in Nature, that furs of animals in general, in thofe countries which have a long and fevere winter, at that feafon change their colour; thus, in Siberia the Fox, &c. are white, and among others the Stoat or Ermine alfo undergoes the fame change; from a fkin of brown and dark yellow, thinly covered, is produced that thick fett, pure white fur which is prized by all, and bought at a very high price by many nations. The procefs or manner how Nature accomplifhes this metamorphofis, is not fo eafily traced:

we well know her maternal care does accomplifh it, as a means of protecting the life of her offspring by a famenefs of colour with the fnow, which for months covers the ground where this little animal is placed to endure the rigours of fevere froft; and, although their efcape is thus facilitated, the wily arts of man fucceed in the deftruction of thoufands, by means of traps of various kinds baited with flefh. The hunters of Norway fhoot them with blunt arrows.

In very fevere feafons the Stoat in England (and very frequently in the further parts of Scotland) is known to change white; but their fkins are of little value, the feverity of our winters not being fufficient to effect fo complete a change in the colour and fubftance of the fur, in which its greateft merit confifts.

An Ermine brought to England quite white in the month of May, loft all its fplendour in about thirty days (beginning at the head), which did never return; its food is rabbits, birds, mice, &c. is very quick of motion, and has a foetid fmell, as all of the Weafel Tribe have.

ANIMAL of the WEASEL-KIND.

OF the limited knowledge of man, and the unlimited bounds of the animal creation, another inftance occurs in the fubject we are now to defcribe. This animal muft be confidered as a non-defcript; the regularity and proportion of the features are fuch, that it appears a perfect animal, that is not a variation produced by chance, nor the off-fpring of a mifcellaneous copulation.

The body is twelve inches long, and with the legs and head is black; on the back are four longitudinal, broad, waved ftripes of white; in the front of the forehead is a fmall triangular fpot of the fame; the ears are fhort and round, and white on the infide, which is continued a little down the face; the tail is 10 inches long, very full and bufhy, the hair foft and fine; the tail in general is down, but erect, as fhewn in the drawing, when pleafed or frightened; has five claws on the fore legs, and four on the hind ones; teeth are very fmall and fine; fleeps in the day time, at prefent in a lady's fkin muff; when awake in continual motion; is very agile and frolickfome; is very tame and docile; anfwers to the name of "Jack," and readily comes to any one when called: legs rather fhort.

Was brought from Bengal laft fummer, by Capt. Gell, of one of the King's fhips; is now in the poffeffion of Sir Jofeph Banks, by whofe permiffion this drawing was made.

Upon the whole, the tail of the animal, when erect, is like that of the Squirrel; but, from the fhape of the body, we think it more properly claffes with the Weafel tribe: indeed, the name handed to us was

"THE STRIPED POLE CAT."

THE BROWN-COTI.

The BROWN COTI

I S an animal of the Weaſel tribe. There are ſeveral varieties of the Coti; the one here repreſented is the ſame as Mr. Buffon names,

Le Coati noirâtre.

The Brown Coti has a longiſh head, the ſnout conſiderably elongated, but not ſo much as in one of this ſpecies of animals. The ears are ſhort, tip of the noſe very flexible, and of a purple colour; a light ſtripe from the noſe to the back of the forehead; cheeks almoſt white, with a patch of white above the eyes; feet ſhort and black; eneral colour brown, mixed with nearly black hairs; the fur coarſe

and long; length of the body about three feet from the noſe to the inſertion of the tail, which is about two feet ſix inches more; the neck and body long.

. The Cotis are natives of South America, particularly Brazil and Guiana; feed on fruits, eggs, and poultry. The one from which this drawing was made, was very greedy. Run up trees, or any perpendicular place, very nimbly; eat like a dog, holding the food between the fore paws; but drink by ſuſtion like a pig; are eaſily made tame, and are much inclined to ſleep in the day-time.

THE BADGER

Drawn from Nature & grav'd by R. Blakly by Tho Cotten font.   Nᵒ __ in the Terrace Tottenham Court Road June 1779.

The B A D G E R.

THE Badger, though a native of England, is an animal not very well known; the fhynefs of his temper, and ftrong propenfity to fleep, which foftens the call of appetite, makes his appearance in fearch of prey lefs neceffary and frequent; the night alfo is the feafon for his fearch; when, whether vegetable productions are fatisfactory, is not certain. The depredations in rabbit-warrens, and on young lambs, are frequently laid to the charge of the Badger; on the other hand, the Badger has little or no fpeed, and being much inclined to fleep, will certainly grow fat on lefs nourifhment than more active quadrupeds.

The Badger digs a habitation in the earth with confiderable dexterity; the fore paws being armed with long and ftrong nails, which work with great expedition: the paffage is of a winding form, leading to feveral apartments, but only one entrance; a bed of foft hay and grafs is provided to induce fleep, and fecure a comfortable repofe. This habitation is fo enviable, that the Fox, whofe abilities for burrowing in the ground are inferior to the Badger's, frequently ejects the lawful tenant by laying his fœtid excrement at the mouth of the hole; the Badger being fo cleanly an animal, that the calls of Nature are never obeyed within the apartment,

The female brings three or four cubs in fummer, which are fuckled for fome time, and afterwards are provided with fuch food as her abilities or induftry can procure. Badgers are frequently eaten, and are faid to make good bacon.

The hair of the Badger is very long, coarfe and rough, which gives it a very uncouth and clumfy appearance, and difguifes the true fhape of the limbs; each hair is tinged with three different colours; the roots a dirty white, the middle black, the extremity afh colour or grey, which has produced the well-known faying, "as grey as a Badger." The cheft and belly are very dark, nearly black; a ftripe of the fame alfo extends from the eye to the ear.

The Badger is common to moft northern countries, and are found in fome warmer ones: the Chinefe are very fond of their flefh, which is often an article in their butchers fhops; are hunted by night for the fake of the fkin; when attacked by the dogs, defend themfelves with great courage, and bite very feverely; are about two feet and a half long, tail about nine inches; have fmall eyes, and fhort round ears.

THE OTTER.

The O T T E R.

THE gradation from one clafs of beings to another is made by almost imperceptible degrees. The amphibious nature of fome
quadrupeds join them in clofe connection with the fish tribe; to which
clafs, the links of the chain fo gradually diminish, that we fcarce know
where to fix its termination. The Otter and Beaver are calculated to
live on land, yet their propenfity leads them to the water; and Nature
has provided fuitably for their destination. Thefe animals have four
feet, and the Beaver has a tail covered with fcales; the gradation then
defcends to the Seal, whofe hind feet anfwer more the appearance and
purpofe of fins; next the Walrus, &c. till all distinction is loft in a complete inhabitant of the limpid fluid.

The Otter is a very voracious animal, eats much, and deftroys infi-
nitely more; for, not content with fufficient to fatisfy hunger, it kills
through wantonnefs, and deftroys for mere victory. On the brink of
fome lake or river, under the bank, where the waves have formed an
excavation, in a gallery of this kind it makes its abode; and, when
purfued, evades the fearch by plunging into the water many yards
diftant from the place where it was expected to be found. In a running
ftream the Otter always purfues its prey againft the current; it frequently finks to the bottom, and any fish paffing over is fure to become
its immediate prey. In ftanding water it hunts them into fome creek,
where they rarely efcape its voracity or cruelty; for it will continue
the hunt for hours, and dragging the prey on fhore, leave it as trophies
of fuccefs. In a few nights, one has been known to deftroy all the fish
in a large pond; will fcent or wind the fish at a great diftance.

The Otter, when taken young, may be tamed, and taught to follow
like a dog, and may even be accuftomed to fish for, and at the command of, his mafter: this perfection of education requires much per-
feverance, but is very profitable when attained. The Otter brings
three or four young at a time; the old ones are rarely or never taken
alive; the hunting them is efteemed good fport, as they fuftain a long
chace, fight boldly, and bite cruelly; indeed, few dogs will venture
to attack them alone.

The colour of this animal is various; in general, a light brown;
neck, cheft, and belly white; the hair rather coarfe; neck long and
thick; head round; eyes very clofe together; ears fmall; tail thickifh,
feet fhort, but very ftrong and flexible, and the articulations fo loofe,
that they can be turned quite round, and brought on a line with the
body; a membrane joins the toes of all the feet; about 2 feet long;
tail 12 or 16 inches; are natives of moft temperate parts of the world;
and are found as far north as Kamfchatka: the fkins are efteemed very
ferviceable for gloves.

THE BEAVER.

N. D. & Jones London Court Road, Jan 1, 1800.

The BEAVER

IS an amphibious animal; and, where the intrusion of man does not prevent, live together in a state of civil government, and appear the only instance of brutes forming a regular community, governed by domestic laws. The time of assembling is about the months of June or July, when a society is formed, which lasts the greater part of the year; the resort is from all quarters, and sometimes a troop of 200 or 300 assemble; the place of rendezvous is generally suitable for the colony, either on the banks of a lake, or on a running water. In the latter case, to guard against a sudden swell of the river, a bank or dam is formed across the stream, frequently of an hundred feet long; this is done by first driving stakes five or six feet long, placed on a row, with small twigs interwoven, and the interstices filled with clay; this dam is 10 or 12 feet thick at the base, and gradually diminishes to 2 or 3 at the top. The side next the head of water is sloped, the other is perpendicular.

The dam or mole being finished, the next care is to erect the several apartments or dwellings, which they build on piles or flakes drove into the ground for that purpose, and are either round or oval, divided into stories, to secure a retreat from swelling floods. The first is below the level of the dam, and is usually full of water; the walls are about 2 feet thick, made of earth, stones and sticks, most artfully laid together; the inside is neatly plaistered as with a trowel. Each house, which is about 8 feet above the water, has two openings, one into the water, the other towards the land. The size of the dwelling is proportioned to the number of Beavers which are to inhabit it; usually from 10 to 30. It has been observed, that 400 Beavers have resided in one large mansion, divided into a vast number of apartments, that had a free communication one with another. These works are finished by August or September; when they begin to lay in their stores, which consists principally of the wood of the birch, the plane, &c. which they sleep in water, in quantities proper for use; the summer food is fresh leaves and fruits; are not fond of fish.

The benefits resulting from patient perseverance have become proverbial, and a more striking instance of the good effects cannot be given than the completion of these surprising works, which are begun by mere instinct, and are finished by mere industry. In the labours of this society every Beaver bears a part; some, by gnawing with their teeth, fell trees of great size, to serve as beams or piles; others drive them along the water, and, with their feet, scoop holes in which to place them; while others help to rear them up. Another party is employed in collecting twigs to weave between the flakes; a third in collecting earth, stones, and clay; while a fourth is busied in beating or tempering the mortar, which is done with the tail; others are employed in carrying it on the broad part of the tail to proper places, and with the same instrument ram it between the piles, or plaister the houses. A certain number of smart strokes given with the tail, is a signal given by the overseer for repairing to certain places to mend any defect, or at the approach of some enemy; and the whole society attend with the readiest assiduity.

The teeth of the Beaver are admirably adapted for cutting timber, or stripping the bark, to which purposes they are so frequently applied. Is an inoffensive animal, and seeks safety rather in flight than conflict. The fur of the Beaver is of great service in the hat trade; it also produces a valuable drug, called castoreum; are hunted, and taken in traps and snares; inhabit most northern climes, but are no where found in such abundance as in Canada in America. The trade for Beavers furs with the Indians is a source of great wealth to the Hudson's Bay Company. The length of the Beaver is about 2 feet, height 1; tail 4 inches broad, 1 or 1½ thick. The colour a fine chesnut brown; the hair of different lengths and finenesss; is the only animal whole toes on the hind feet are joined by a membrane, while those on the fore feet are not; the front feet supply the place of hands, similar to the Squirrel.

The G L U T T O N.

THIS singular animal, on account of the length of body, and short-ness of legs, appears to belong to the Weasel tribe. Mr. Pennant allots it with the Bear—it has a roundish head, with a blunt nose, short ears, limbs large and strong, tail very bushy, general colour black, with a broad horizontal stripe, of a yellowish colour, along the upper part of the face, the sides, and the tail.

The great voracity of this animal has fixed upon it the opprobrious name it bears. If the active speed which wild animals in general possess fell to the share of the Glutton, he must inevitably soon thin the forest of its inhabitants, but the cautious hand of Nature has guarded against his voracity by a body ill-formed for celerity; thus disqualified for pursuit, yet ever pressed by an active appetite, it has recourse to cunning and stratagem. Selecting a tree whose situation is promising, or observ-ing on the bark the marks of the teeth or horns of the deer, or other beast, he readily ascends, and, hiding among the spreading branches, he will there wait for weeks together, expecting some unwary animal to pass under, which be instantly drops upon, fixing his teeth and claws into the neck, digs a passage to the great blood vessels, which lie in those parts—in vain the tortured animal flies for relief among the branches of the forest, the Glutton still holds his station; and, although it often loses parts of its skin and flesh, which are rubbed off against the trees, yet it still flicks fast, the force of appetite and nature prevail more than his feel-ings; and he never seizes, but he brings down his prey, wearied by fatigue, and faint by loss of blood: the moment of victory rewards for former trouble, and he then makes up for past fatigue by imme-diately falling to, and ceases not, till overgorging has destroyed every animal function; thus torpid through satiety, he lies till nature qualifies him to renew the feast, which he does not quit till entirely eaten up bones and all. As such a bountiful repast cannot always supply his voracity, he uses much cunning to procure his prey; he will frequently anticipate the sportsman by clearing his traps of the game; he steals upon the retreat of other animals, particularly the rein deer, of whose flesh he is greedily fond: he also lies in wait, and falls upon the game other animals have run down, his constant necessities producing a pretty fertile invention. One of these animals confined at Dres-den consumed thirteen pounds of flesh every day, and yet not satisfied. The Glutton inhabits the northern parts of Europe, Siberia, and Ame-rica; its skin is highly esteemed for a beautiful gloss and damasked appearance; in length it is about three feet and a half, and eighteen or twenty inches high.

THE ARMADILLO.

The ARMADILLO.

WHEN we speak of a quadruped, imagination reprefents an animal covered with hair; as when we mention a bird, or a fish, to the one we attribute feathers, to the other scales; and these distinctions, at the first, appear to mark the boundary of each species; yet nature, as if in defiance of rule, and wishing to astonish as much by particular exceptions as by general laws, so blends her several productions, that it is no easy matter to draw a distinguishing line, and say to which class an animal, whose tail is covered with scales, belongs; or of which family one inclosed in large scales or shells is a part. It therefore becomes us not to judge by one character only, which so often is incomplete.

The Armadillo is one instance among several of a quadruped covered not with hair, but with a shell or shells. Of this animal there are several kinds, whose variety consists in the number of the bands of shell which encircle or cover them; to some the incrustation is divided into only three distinct pieces, to other into six, eight, nine, twelve, and eighteen pieces; which have been considered by some as marks of age; but, in general, with more propriety, have been regarded as different kinds.

The bands of shell lap over one another, and are united by a membrane, like the shell to the tail of a lobster; this shell, or combination of shells, covers the head, the upper part of the body, and the tail; the throat and belly being the only parts not secured: this deficiency is provided against by the power the Armadillo has of rolling itself up like a ball, and thereby covering the vulnerable parts. In time of

danger, when it cannot make good its retreat to its hole, it brings the head and feet close to the belly, and, bending the back, forms nearly a sphere, the tail laps over the joining, and makes a firmer hold; in this form it defies the attack of any quadruped, and a patient suffering of insult generally proves its security; but man, whose power is over the whole creation, whose power and perseverance is irresistible when any good is to be obtained, or any luxury enjoyed, soon convinces the poor Armadillo of its danger, by exposing it to the fire, which makes it quickly unroll.

The small kind of this animal are esteemed very nice eating, and are therefore hunted with avidity; dogs are used to pursue them, who impede their flight by making them roll themselves up, when they become an easy prey to man. As they run pretty fast, if a few minutes are allowed, they immediately fall to work, and seek security by burrowing in the ground, which they do with great celerity, and must then be dug out. Their accustomed abode is in holes of considerable depth, and, as they wander only by night, and then not far, some industry is required in securing them.

The colour of the shell of the Armadillo is a greyish yellow; that part of the head which is not covered, is a blackish brown, the belly a yellowish white, which bears evident marks of a tendency to ossify; the feet a fleshy red colour, are spotted. Are natives of South America, particularly the Brasils; about 14 or 18 inches long; the larger kind 2 feet.

THE BOMBAY SQUIRREL.

London, from the Original Drawing Publish'd by Cha Catton Jun.

Engr. on the Stone. Wholesale Court Road, Jan. 1730.

The BOMBAY SQUIRREL.

THE task of the Historian or Naturalist is often surrounded with perplexities and difficulties; sometimes, from the incertitude, the variation, and almost total dissimilarity of his information; and sometimes from the entire want of every information; for, of those who are pleased with possessing a new or strange animal, and will be at the trouble of transporting it from distant parts, few have abilities or inclination, to seek that knowledge which would satisfy a Zoologist in its habits and propensities; and, perhaps, very few enjoy opportunities of acquiring certain intelligence.

With respect to the animal now before us, we acknowledge our want of certain information; all we can say is, that it was brought from India in one of the Company's ships; it appeared to have all the motions and actions of a common Squirrel; and, as its size was larger, so its strength was greater. The fine rich colouring of the fur gave it a very grave and majestic appearance. We are sorry to add, that since this drawing was made the animal is dead.

The length of the body was 15 inches; the tail as long; the head longish and round; the ears tufted; the colour of the head and ears a fine deep brown; the shoulders, along the back, hams and tail, black; sides a reddish purple; chest, fore feet, belly, and inside of hind feet, a yellowish white. The end of the tail to this animal was not of an orange colour, as the one described by Mr. Pennant was,

THE PECCARY.

The P E C C A R Y,

AT first view, bears a general resemblance to our common hog, but on examination is evidently of a distinct species; neither will they breed together. The head of the Peccary is large; the snout long, and terminates like the hog's; the neck is thick and short; the body bulky, and marked down the neck with a belt of a whitish colour; the legs are short; the general colour is black; each hair or bristle is marked with alternate bands of black and white, like the porcupine's quills; the coating is a coarse kind of bristles, which are long over the whole body, and the length of four or five inches along the back; has no tail; the size is rather smaller than the common hog; the appearance equally clumsy with all of this tribe.

The Peccary is further distinguishable from every other quadruped, by an orifice in the back, near the rump, which by some has been mistaken for the navel; from this opening discharges an ichorous liquor of a disagreeable smell. It is necessary, immediately on killing the animal, to extract this orifice or gut, else, in the course of a quarter of an hour, it will taint the whole carcase.

The Peccary is a native of the hottest parts of South America, where they are very numerous, and go in, herds of two or three hundred; prefer the mountains to the plains, and the woods to the open parts, as the food they most delight in abounds there in the greatest plenty; they eat also toads, lizards and serpents; the latter they skin with great adroitness, holding them with the fore feet.

The Peccary, though not armed with such offensive weapons as the wild boar or hog, will fight stoutly with the beasts of prey. The Jaguar, or American Leopard, is its mortal enemy; often the body of that animal is found with several of these hogs, slain in combat. They render mutual assistance when attacked, and endeavour to surround the enemy. The Peccary may be rendered tame and domestic; is satisfied with the same food as the hog, but is not so much inclined to be fat; nor will it, like them, wallow in mire. The flesh is esteemed very good food.

The PORCUPINE.

THE very fingular properties of this animal have been the foundation of many fabulous reports.

Nature, in all her productions, gives to every animal fome particular quality—in fome fhe implants a fierce and favage cruelty, regarding only the gratification of an inordinate appetite; in others a mind harmlefs and peaceable, yet poffeffed of powers and faculties to act on the defenfive, fuch as extreme caution, or cunning, which may be called an internal defence; or an external one, fuch as fhields the animal now before us—thus guarding the weaker againft the overbearing oppreffion of the ftronger, thereby preferring an equality and balance in her productions. The Porcupine is in its nature quiet, and feldom gives provocation of offence; and when attacked by an animal of prey defends itfelf by erecting its quills in fuch a manner as always to keep them pointed towards the enemy; thus fecuring its own fafety. Sparman, a modern traveller to the Cape, reports, " By rolling up its body like the " hedge-hog into a heap, and fetting up its prickles, or quills, many of " which are a foot and a half long, it is perfectly well defended from " dogs, as well as other animals."—If time permits, it makes towards, and afcends a tree, where perched in fafety, it wearies the patience of its purfuer.

The power formerly attributed to the porcupine of voluntarily difcharging its quills, and with them mortally wounding, at a confiderable diftance, is now entirely difcredited, great provocation having been ufed, if poffible, to produce this effect, but without fuccefs. The quills are ftrongly inferted in the body of the animal; each one fed or fupported by a fmall ball or nucleus of a foft fpongy matter, varying in fize according to the bulk of the quill.

Thofe Porcupines which inhabit the Eaft are reported to poffefs a Bezoar or ftone, which is reputed an antidote to poifon; this is found in the head, and is confidered of great value. Taverner reports his giving five-hundred crowns for one, which he afterwards changed to advantage; he fays they are alfo found in the belly fometimes. Of this animal there are feveral fpecies little akin otherwife than in being provided with a coat of quills. The one here drawn has the upper lip divided, head like an hare, with a row or ruff of ftiff briftles furrounding it on the forepart of the fhoulders and top of the head, reclining backwards; the body part is thickly covered with quills from nine to twelve inches long, very fharp at the point, and regularly annulated with alternate black and white; fome of the larger quills are near a quarter of an inch diameter; the internal fubftance is fpongy, like the upper part of a goofe-quill; the body is thickly covered with hair between the quills; the head, belly, and legs are covered with ftrong briftles rooted among foft hair; the feet are fhort, as is the tail, which is covered with quills; the general length of the animal is about three feet; it inhabits Africa, India, Tartary, and Paleftine. In Italy a fpecies with fhorter quills run wild; thefe are fold in the market at Rome, where they are eat:—The traveller before quoted fays, " the flefh neareft re-" fembles pork, a circumftance which undoubtedly gave it the name it " bears; it is chiefly ufed as bacon, being fmoked and dried up the " chimney for that purpofe, and is by no means ill-tafted." It feeds on fruits, roots, and herbs; the colour inclines to black.

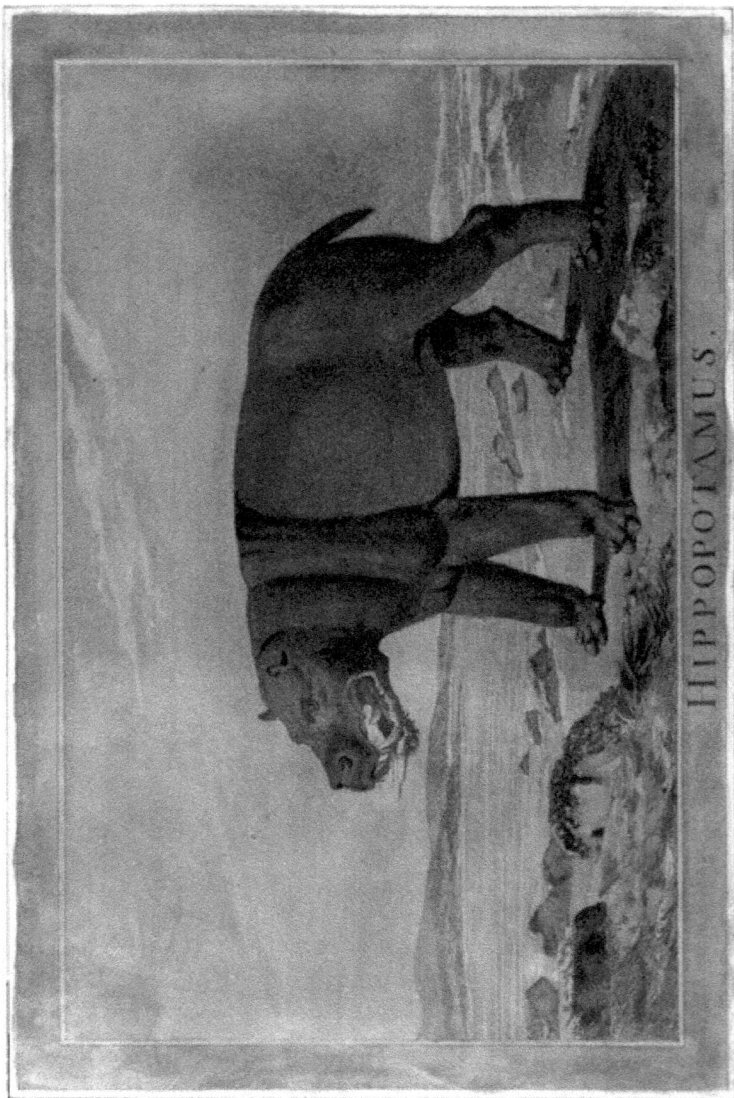

HIPPOPOTAMUS.

Drawn from Life Engraved & Published by The: Cooton Junr. & St: c/a & John's street)

No 1 in the Toyman London Exuet. Pupid Aug. 18. 1787.

The HIPPOPOTAMUS.

THIS is an amphibious animal, of a gigantic bulk, inferior only to the Elephant; is found in large and rapid rivers; the Nile, the Gambia, and the great rivers on the South-Eaſt parts of Africa. The general dimenſions of this animal are reported as follows; in length about 17 feet, circumference 15, height 7, legs 3, head 3½ feet long, and 9 round; tail thoriſh, eyes and ears ſmall; the latter pointed, covered on the inſide with hair; the two tuſks in the lower jaw lie nearly horizontal, and meaſure about 27 inches long; their weight about 3½ pounds: theſe tuſks are highly prized by Dentiſts, for making falſe teeth, not being ſo liable to turn yellow as ivory. The ſkin thinly interſperſed with ſtrong hairs, or briſtles; is very rough, of a mud colour, and, when newly riſen out of the water, the animal has a glittering or ſlimy appearance.

This animal, unleſs inſulted, is of a quiet diſpoſition; paſſing its time in wandering up and down the rivers, which it chiefly inhabits. Fiſh, as ſome have reported, makes no part of its food; this it ſeeks at night on land, feeding on the graſs, reeds, and boughs of trees which are in the neighbourhood of its abode. "The quantity of graſs," ſays Mr. Sparman, "which I have at different times obſerved to have been confuſed "by one of theſe animals, in ſpots where it has come over night to "graze, is almoſt incredible." Indeed, conſidering the bulk of its body, and the great ſize of its ſtomachs (which are four), it certainly muſt require great quantity of nutriment. When in the water, the Hippopotamus frequently riſes to the top to take in air; it will ſuffer immerſion about 30 or 40 minutes. Providence, in its univerſal wiſdom, has appointed the abode of this coloſſal animal in very diſtant parts from the habitations of man, elſe its great ſtrength and revengeful nature muſt produce much miſchief;——of its ſtrength, it has the credit of being able, with eaſe, to bite a man in two; and of its revengeful nature, it has been known to place itſelf under a boat, and by riſing up, overſet it with ſix men in. Moore, in his travels up the Gambia, relates a ſimilar diſaſter: A boat going down the river, fell in with a herd of theſe animals: "on being fired at," ſays the Narrator, "before the flaſhing of "the pan was well out of our eyes, being in the midſt of them, one "which we ſuppoſed was wounded, flounced and kicked about the "boat till he knocked a piece out of the bottom; and before we could "reach ſhore, the ſunk right down." One of theſe animals purſued for ſeveral hours a Hottentot, who found it very difficult to make his eſcape; their aſtivity when on land muſt not, therefore, be calculated from the unwieldineſs of their appearance. In the water they ſwim with great vigour againſt the ſtrongeſt current, and frequently ſunk to the bottom, and walk as on land: they frequent ſalt water, but do not drink it; when angry, make a furious noiſe, between the grunt of a hog and the neighing of a horſe, which, probably, fixed on it the name of Hippopotamus, which is Greek, and ſignifies a river horſe; its number of ſtomachs has certainly cauſed it to be called the ſea cow.

A conſiderable portion of a ſkeleton of one was lately found in digging at Chatham, which has led a learned Gentleman to make ſome ingenious queries concerning the antiquity of the earth, the climate of this country in former times, and to conclude this animal was once a native of England. The Hippopotamus is certainly the animal which is deſcribed in ſuch a figurative, yet correct, manner in the 40th chapter of Job.

The drawing for this ſubject was taken from a ſtuffed ſkin in the Leverean Muſeum, and is regarded as a juſt figure of the animal, though not at full growth; the dimenſions were 9 feet long, and 5½ high.

CROCODILE.

Drawn from Life Engraved & Published by Cath. Waters for this Work.

N.º 12. the Town London Road Augt. 1st. 1789

The CROCODILE.

THIS amphibious animal is claffed among the Lizard Tribe. A courage fierce and favage, aided by great bodily ftrength, joined to a confiderable fhare of cunning, or ftratagem, compofe the great outline of this animal's character.

Of a bulk truly formidable from 18 to 28 feet long, they are univerfally dreaded; always on the watch, with activity and appetite ever ready, the Crocodile lets flip no opportunity of committing his depredations on animal nature; the water is his proper element, but if his voracity has caufed a fcarcity of game here, hid among the reeds, he lies in wait on the banks of the river, expecting the approach of fome thirfty animal, compelled by the heat to regale nature with a lap of water, then the Crocodile immediately feizes upon, and pulls down his prey; where, unlefs of very large bulk, it rarely efcapes being prefently drowned, holding his prey both by his claws and his mouth, which, in one 17 feet long, will open near 2 feet, a gape fufficient to take firm hold of man or beaft.

This animal is oviparous, or generated by eggs, which the female depofits with the utmoft fecrecy and circumfpection in the fand, on the fhore of the river; fcratching a hole in a fuitable place, in about an hour fhe depofits near an hundred eggs, then covers the place with the moft fedulous anxiety for their fafety; the fame tafk is performed the fucceeding and third day, when about 300 eggs are depofited, thefe covered with great care with the fand, fhe commits to the foftering hand of Nature: the heat of the fun in about 30 days animates the eggs, and now Nature prompts the mother to feek after her young by clearing away the fand; the brood thus liberated, fome take immediately to the

water, while others mounted on her back, are introduced to their fluid habitation with more eafe and fafety; this parental care foon fubfides; their proper element once gained, fafety depends on their agility and caution.

The ferocity of the Crocodile, like other wild animals, very much abates as his abode is more or lefs in an inhabited country. In unfrequented rivers they lie bafking in droves together, and have the appearance of large trunks of trees, with rough and rugged bark floating on the water; yet, thus apparently torpid, appetite is awake, and the approach of any animal is quickly followed by a conflict for victory. In more populous countries the undivided tyranny of man has reduced them within better bounds. It is reported, the children of the Siamefe play with them in a very familiar manner, and will even correct them with blows; it is true, indeed, thefe people treat them more as friends than enemies. The reverfe to this is the general character of the Crocodile, whofe great fecundity muft be very alarming, had not the wife, the beneficent hand of Providence appointed a bird of the Vulture kind, and an animal called the Ichneumon, with an appetite peculiarly fond of the Crocodile's eggs.

The colour is a greenifh brown, the upper part of the body is covered with a very thick and rough fkin, proof againft the edge of a fword; the belly, of a greenifh white, is more vulnerable; the eye is very prominent and large, of a yellowifh green; the feet are fhort, but very mufcular; with its tail it ufually knocks down and ftuns its prey. Inhabits moft great rivers of Afia, Africa, and America: the Nile in Egypt has ever been famous for them.